Nature's
Partners

BY JOAQUIN CARR

TABLE OF CONTENTS

Chapter 1

Working Together

One night, your brother comes to you with a problem. It is his turn to wash the dishes. But he has a math test tomorrow. He needs to start studying right away. He asks you for a favor. Would you wash the dishes tonight so he can study?

You wash the dishes as a favor to your brother.

Your brother puts a program on your computer as a favor to you.

You agree to help him. Then you remember you need to ask a favor, too. You have just bought a new program for your computer. You want to use it this weekend. Your brother is good with computers. You ask him if he will put the program on your computer tomorrow.

He says he will. You two have made an agreement. You will help each other. You will also both get something in return. This is a good partnership.

Chapter 2

Nature's Partners

There are many partnerships in nature. Sometimes they are between two different **species,** or kinds of animals. These partnerships help one or both partners. Scientists have a name for this relationship between two living things. It is **symbiosis.**

Symbiosis can exist between plants and animals. This bee will carry the pollen from this flower to another flower.

Symbiosis takes place when two **organisms** live
in a way that helps at least one of them. In some ca
partners help each other. In other cases, one partner
out the other is not. Yet the other partner is not har
still another kind of symbiosis, one partner is helped
harming the other partner in the process. In this boo
you will learn about nature's partners. These
partnerships are all types
of symbiosis.

Redbilled oxpeckers
feed on parasites on the
mpala. This is an example
of symbiosis.

Mutualism	Commensalism	Parasitism
In mutualism, both partners help each other.	In commensalism, only one partner is helped. The other is not harmed.	In parasitism, one partner is helped, but the other is harmed.

There are three main kinds of symbiosis. The first is called **mutualism**. In mutualism, both species need each other. They become partners. The partners help each other.

Sometimes symbiosis does not help both partners. One partner gets what it needs. The other partner does not get anything in return. But this partner is not harmed, so it lets the relationship continue. This second type of symbiosis is called **commensalism**.

The third type of symbiosis is called **parasitism**. Only one partner gets what it needs. But it gets what it needs by harming its partner!

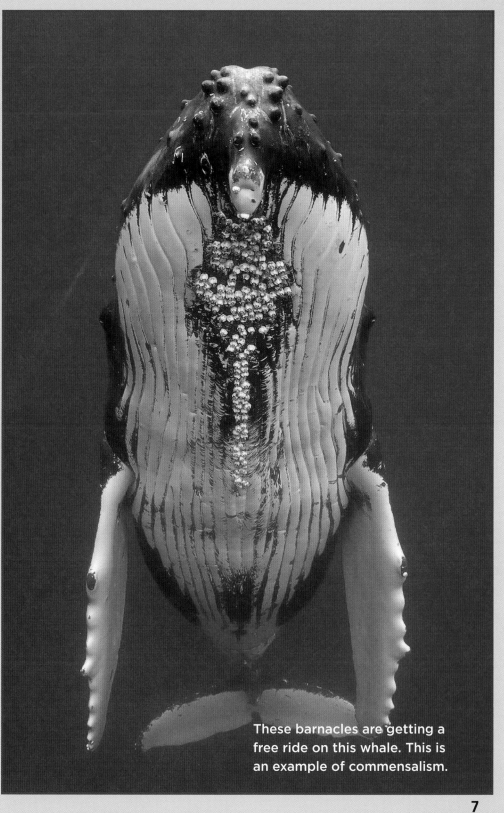

These barnacles are getting a free ride on this whale. This is an example of commensalism.

Chapter 3

Mutualism

You can find examples of symbiosis everywhere in nature. Let's look at mutualism first. It takes place between two different species. These species work side by side. Each helps the other reach a goal. Let's start with the ocean.

Goby fish and shrimp are good roommates. They live on the ocean floor. Gobies need a safe place to hide from enemies. Shrimp give the gobies a safe home. Shrimp dig holes in the sand. The two live together in these holes.

Why do these shrimp share their little caves? Shrimp are nearly blind. They cannot see well enough to look for food. Gobies bring back food for themselves and the shrimp. They also warn shrimp when enemies are near. Gobies tap the shrimp with their tails to warn the shrimp to take shelter in the hole.

The shrimp and the goby fish is an example of mutualism. Both partners are helped.

There are more examples of mutualism in the ocean. Some small fish clean the bodies of larger fish. One of these tiny fish is the wrasse.

Wrasses act like little underwater dentists. They clean the teeth of larger fish and get a meal along the way. Wrasses scrape the **parasites** from the teeth of barracudas, sharks, and other large fish. In fact, they swim right into the big fish's mouth!

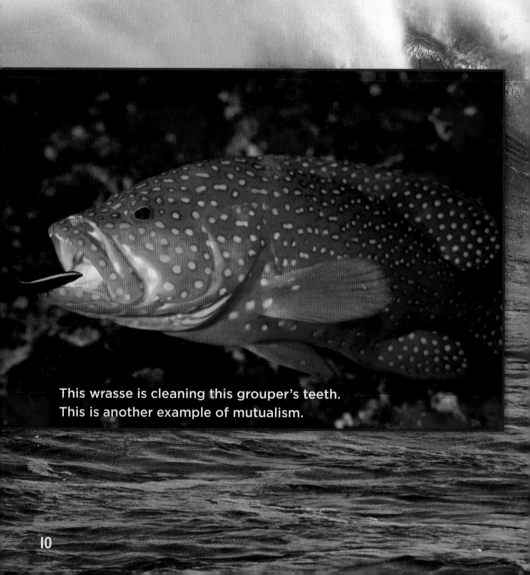

This wrasse is cleaning this grouper's teeth.
This is another example of mutualism.

What do the wrasses get from this partnership? They get food! Parasites are a meal for the wrasses. Of course, the larger fish don't mind getting rid of the parasites. Parasites can harm them.

Some big fish even try to find the wrasses. They will wait in long lines for a teeth cleaning. They roll over on their sides when it is their turn. This teeth cleaning keeps them healthy.

This wrasse is cleaning the parasites from the teeth of this moray eel. The moray eel gets a teeth cleaning. The wrasse gets to eat the parasites.

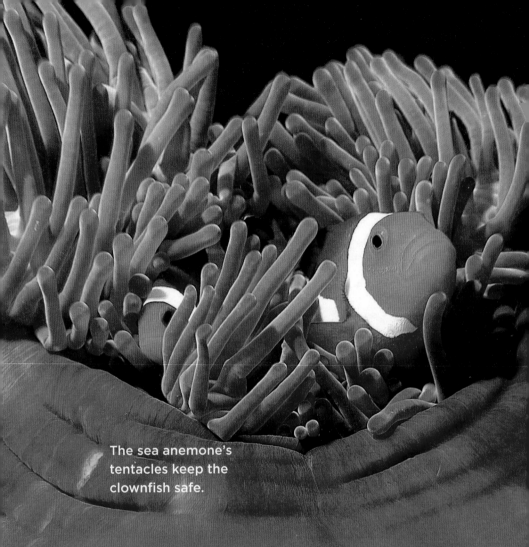

The sea anemone's tentacles keep the clownfish safe.

The clownfish and the sea anemone (uh-NE-muh-nee) are partners, too. Clownfish have bright colors. This makes them easy to see. Sometimes, bigger fish come after them. They chase them right into the sea anemone's "arms," or **tentacles**.

The clownfish hides in the sea anemone's tentacles. The tentacles keep the clownfish safe. Anemones also have special stinging cells. Their tentacles can stun larger fish. Then the anemone eats the fish. Both partners get what they need. The clownfish stays safe. The sea anemone gets a meal.

Perhaps no partner takes a bigger risk than the Egyptian plover. The tiny plover is one brave bird!

Plovers are partners with crocodiles. They even fly into a crocodile's mouth! Once inside, they eat leeches that have clamped onto the crocodile's gums. Leeches suck the crocodile's blood. Of course, the crocodile doesn't mind when the plovers eat the leeches. Crocodiles even wait for plovers with their mouths wide open!

Plovers eat leeches right out of the crocodile's mouth!

Ants and acacia bushes have
an interesting partnership.

As you know, symbiosis does not always take place
between two animals. Sometimes a plant and an animal are
partners. Some ants team with acacia bushes. These bushes
have a sweet nectar. The nectar is a juicy treat for ants.

The ants protect the acacia. They will attack anything that nibbles on the bush and harms it. This includes grasshoppers and aphids. The ants will even bite nibbling cows!

It is no wonder the ants are such great guards. The acacia bush gives them both food and shelter.

Ants protect the acacia bush. In turn, the acacia bush gives the ants food and shelter.

Some partners need the same thing. But they cannot get it on their own. That is the case with the honey badger (or ratel) and the honeyguide. What do these animals have in common? You guessed it. They both like to eat honey!

The honeyguide is a small African bird. It is great at finding bees' nests. But there is a problem. The honeyguide's beak is not strong enough to break open the nests. They cannot get to the honey. Once again, symbiosis saves the day!

This honeyguide has found a bee's nest.

When a honeyguide finds a bees' nest, it calls out to its partner—the honey badger. The badger loves honey, but it has trouble finding nests. It needs the honeyguide's help.

Once the badger arrives, it rips open the nest with its sharp claws. Then the honey badger eats the beeswax. After eating its share, the honey badger leaves. The honeyguide then eats the leftovers.

The honey badger needs the honeyguide to find the bees' nest. But the honeyguide needs the honey badger to break open the nest. Both partners get to eat the beeswax.

Chapter 4

Commensalism

Commensalism is when one partner is helped without harming the other. The partnership between the remora fish and the shark is a good example of commensalism. Remoras have a strange feature. They have flattened suction cups on the top of their head!

The remora and the shark is an example of commensalism. The remora is helped by the shark, but the shark is neither helped nor harmed.

Remoras use this feature for their own gain. They stick their suction cup on sharks. Then they go for a ride! Sharks pull them along. Remoras must feel safe. After all, few sea creatures will attack a shark. Remoras even get shark leftovers. They eat any scraps left floating in the water. Sharks do not benefit from remoras. But they are not harmed, either.

Parasitism

Parasitism is another example of symbiosis. But this relationship is one-sided. One partner gives the other food or a home. The partner that provides for the other is called the **host**. The other partner is the parasite. It is the guest. But parasites are bad guests. They take, but they do not give anything back. In fact, they even harm their host!

The dodder plant is a parasite. It takes away the food and energy that the host plant needs.

Fleas are parasites. They harm their host.

A veterinarian examines this dog for fleas.

There are many kinds of parasites. Fleas harm their host. Their host might be a dog or a cat. Ticks are parasites, too. They dig beneath the skin of their host. Then they suck their host's blood. Some plants are parasites, too. The dodder is one example. It grows around its host plant's stem. It depends completely on its host for food.

Chapter 6

Living in Harmony

Symbiosis is an amazing part of nature. Some living things stay with their partners for life. Other partners stay together for a shorter time. Then they find new partners.

How do partners find each other? For example, how does the honey badger and the honeyguide know to team up? Scientists do not have all the answers. What they do believe is that partners find each other in order to survive.

Hermit crabs and sea anemones are partners for life.

Glossary

commensalism (kuh-MEN-suh-liz-uhm) a relationship between two living things that helps one partner without harming the other *(page 6)*

host (HOHST) the organism a parasite lives in or on and harms *(page 20)*

mutualism (MYEW-chew-uh-liz-uhm) a relationship between two living things that helps both partners *(page 6)*

organism (OR-guh-niz-uhm) a living thing *(page 5)*

parasite (PAR-uh-site) an organism that benefits from living in or on another organism *(page 10)*

parasitism (par-uh-sit-IZ-uhm) type of symbiosis where one organism is helped while the other is harmed *(page 6)*

species (SPEE-sheez) a group of animals or plants that has many of the same features *(page 4)*

symbiosis (sim-bye-OH-sis) a relationship between two living things that helps at least one of them *(page 4)*

tentacle (TEN-tuh-kuhl) a long, thin body part that helps an animal feel, grasp, and move *(page 12)*

Index